YOUR KNOWLEDGE HAS VALUE

- We will publish your bachelor's and
  master's thesis, essays and papers

- Your own eBook and book -
  sold worldwide in all relevant shops

- Earn money with each sale

Upload your text at www.GRIN.com
and publish for free

Bibliographic information published by the German National Library:

The German National Library lists this publication in the National Bibliography; detailed bibliographic data are available on the Internet at http://dnb.dnb.de .

Imprint:

Copyright © 2015 GRIN Verlag, Open Publishing GmbH
Print and binding: Books on Demand GmbH, Norderstedt Germany
ISBN: 9783668151444

This book at GRIN:

http://www.grin.com/en/e-book/315193/a-survey-on-the-two-factor-authentication-protocol-used-in-the-telecare

Kishan Makadia, Nishant Doshi, Sunil Vithlani

A Survey on the Two Factor Authentication Protocol used in the Telecare Medical Information System, including possible attack scenarios

GRIN Publishing

A SURVEY ON TWO FACTOR AUTHENTICATION PROTOCOL USED IN TELECARE MEDICAL INFORMATION SYSTEM, INCLUDING POSSIBLE ATTACK SCENARIOS

Kishan Makadia, Sunil Vithlani, Nishant Doshi

A Survey on Two Factor Authentication Protocol used in Telecare Medical Information System, Including Possible Attack Scenarios

Submitted By
KISHAN NAVINBHAI MAKADIA

Under the Guidance of
SUNIL VITHLANI and NISHANT DOSHI
Department of Computer Engineering
Marwadi Education Foundation Group of Institutions

A Dissertation Phase-1 * Submitted to

Computer Engineering Department
Marwadi Education Foundation Group of Institutions
Affiliated to Gujarat Technological University

For the Partial Fulfillment of Degree of Master of Engineering
In Computer Engineering

December – 2015

Faculty of P.G. Studies & Res. in Engineering & Technology

At & PO : Gauridad, Rajkot-Morbi Road

Rajkot 360 003. Gujarat. India.

*Report is modified and update based on Dissertation Phase-1 Presentation and

comments

Index

List of Figures

List of Tables

Acknowledgements

I am grateful to numerous local and global "peers" who have contributed towards shaping this Dissertation Phase 1 report.

First and foremost I would like to express my sincere thanks to *Dr. Nishant Doshi* who created my interest in *cryptography*. He was always there to guide, motivate and support me whenever I was stuck. He constantly reminded me to achieve my goal. His observations and comments helped me to establish the overall direction of the research and to move forward with investigation in depth. Irrespective of his busy schedule, he always gave time to listen to my doubts patiently and gave valuable suggestions like a parent. His doors were always open to discuss my doubts anytime.

I owe my deep sense of gratitude to Dissertation Phase 1 report examiners *Dr. Sarang Pande, Dr. Nitul Dutta* for their valuable suggestions and critical comments during presentation of credit as well as progress seminars and also sharing their knowledge which influenced me more to carry out this research work. I am thankful to *Krupali Dalsania*, for helping us to solve typos and grammatical error throughout the book.

I thank to all my student colleagues for providing fun filled and very informative environment to learn and grow. It is their love and encouragement, which helped me a lot during my research work. I had many memorable moments with them inside and out-side my work. I am grateful to all my *colleagues, M.Tech students, Teaching Assistants* and many others, for being with me in my difficult times and for all the emotional support, care, and fun they provided and who have been kind enough to advise and help in their respective roles. I owe a dept of gratitude to all my friends for their guidance and support. *Manish Shingala and Mayur Oza* for their thoughtful discussion related to research work that helped me to complete this work in timely fashion. *Ruchita Kaneria, Harshit Champaneri, Dhara Patoliya and Jinita Tamboli* for constant motivation to carry out this research work.

I wish to thank staff of *Department of Computer Engineering, MEFGI* for providing me resources throughout my stay in the college.

Last and most important, I thank my family members. Without their constant support, motivation and love, I would not have been reached so far. I dedicate this research work to my family.

Kishan Makadia

Abstract

Since last few decades, there is drastic increase in the availability of lower-cost communications systems and healthcare services. The telecare medical information system supports health-care delivery services. These systems are moving a digital world, where automatic patient medical records interconnected telecare medical information system. It is important to assurance the privacy and the security of the patient in the telecare medical information system. A secure authentication scheme will provide to data integrity, availability and confidentiality. Authentication scheme is required the guarantying the security of users in Telecare Medical Information System. Authentication is used to verify the validity of the users and Telecare Medical Information System server during remote access.

Keyword. Telecare medical information system, User authentication, Security, Password, AVISPA.

1 Introduction[1]

In order to protect users' privacy, such as mobile number, medical record number, health information, authentication scheme for telecare medical information systems (TMIS) has been studied widely [26]. An authenticated mechanism for a user is required to protect that the secure information is not received by any illegal persons. The current availability of lower-cost telecommunications system and tradition physiological monitoring devices has made it possible to take advantage of telecare medical right into the patient's home, i.e. a connection between patient's home and doctor at a hospital center or home health-care (HHC) agency [27].

During past there was queue to take a doctor's appointment medical services. But today person, just a click allows to access medical services from anywhere and anytime. Tradition's time consuming methods on medical services has been restore by digitalized smart technique. We have entered in a new global world of error-free, efficient and quality of healthcare services.

It is important to ensure that privacy and security of the patient in the Telecare medicine information system. In the Telecare Medical Information System, the privacy and security problem have patients' rights to understand and protected their private information. Never the less security in the Telecare Medical Information System is of prime concern. The important concern about security problem is how to protect information security and privacy during transmission over Internet.

Smart card based authentication is kind of a two-factor authentication protocol, firstly a successful authentication which need to the user have a valid smart card and password.

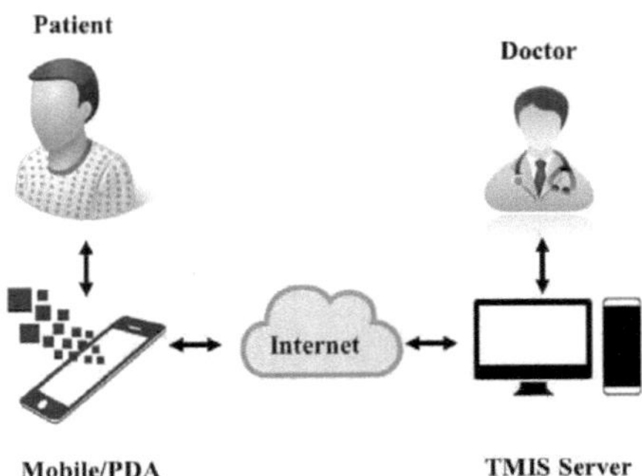

Fig 1.1 Telecare medical information system overview

---

[1] The content of this book is based on the suggestions received in Dissertation Phase I presentation and report towards master degree of Mr. Kishan Makadia.

2 Literature Survey

2. 1 Classification Tree

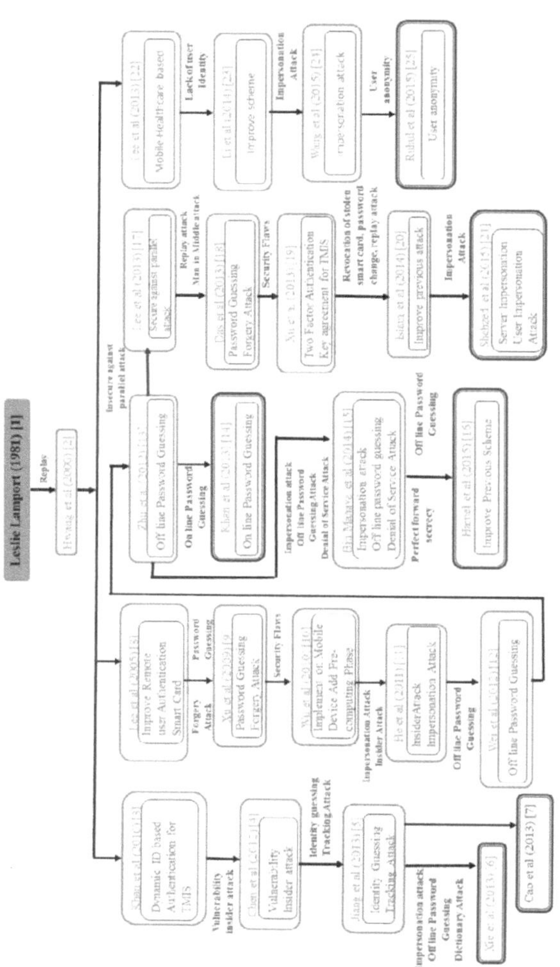

Fig 2.1.1 Survey Tree of TMIS

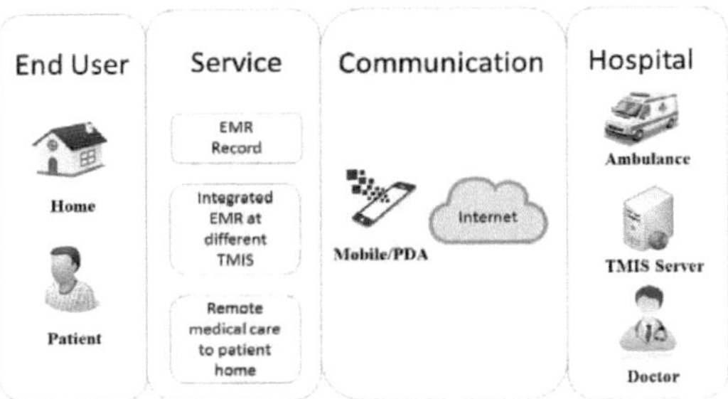

Fig 2.1.2. Framework of telecare medical information system

2.2 Review of different Author's scheme

In 1981, Lamport [1] introduced the first password-based authentication scheme using password tables to authenticate remote users over insecure network. Later on many password authentication schemes [14, 15, 16] has been proposed to improve security or efficiency or cost.

Secure authentication schemes are the services provided by the server which are not accessible to illegal person. While widely execute, such machine will inescapable suffer from various attacks. In order to solve these difficulties and stronger the security of the system, smart card based password authentication schemes are introduced.

As before smart card based password authentication is a type of two-factor authentication scheme and a triumphant authentication requires the user to have a valid smart card and a right password. Few smart card based password authentication [5,6] have been present in the last few decades.

To reach a high level of security and deliver to a secure authentication, there are few of the password based authentication strategy are proposal and presented the idea of Diffie-Hellman key exchange [26], which depends on the issue of discrete logarithm' s problem. The disadvantage of these schemes is that the computation cost is costly which is not suitable for small computation power equipment's in telecare medical information system.

Xu et al. [9] has been proposed the authentication schemes based on discrete logarithm' s problem and smart card. there are some security problem leading their schemes to be unconfident. Wu et al present a systematic password-based authentication scheme. The large contrast between Wu et al scheme [10] and other author's schemes is the addition of pre-computing phase.

The Wu et al scheme in user authentication scheme hire three categories of cryptographic and mathematical technique and theorems, hash function, cryptograph, and DLP.

Wu et al present password-based user authentication scheme is secure and acceptable to be implemented. The differentiation of Wu et al scheme with other related schemes as outline in Table 2.2.1. Xu et al. [9] scheme suffers from unconfident attack.

The performance comparison of Wu et al [10] scheme with the other author schemes is shown in Table. the listed three schemes [9] need more exponential working leading to the need for more computation time resulting into incompetent, while Wu et al [10] scheme present scheme only needs two exponential operations and eight hash function in performing the authentication strategy. The time-consumption of computation time on patient side of this scheme is small than others. This scheme view that the scheme is more systematic and appropriate to collocating with low power gadget for the telecare medical information system.

Table 2.2.1 Comparisons of the security attack

	Yang et al scheme	Liu et al scheme	Chen et al scheme	Xu et al scheme	Wu et al scheme	He et al scheme
Replay attacks	secure	secure	secure	secure	secure	Secure
Password guessing attack	secure	secure	insecure	secure	secure	Secure
Impersonation attack	secure	insecure	insecure	insecure	secure	secure
Stolen-verifier attack	insecure	secure	insecure	secure	secure	secure
Session key security	secure	secure	insecure	Insecure	secure	secure
Perfect forward secrecy	insecure	secure	insecure	secure	Secure	secure
User anonymity	insecure	insecure	insecure	insecure	insecure	insecure
Insider Attack	insecure	insecure	insecure	insecure	insecure	secure

Table 2.2.2 Comparisons of the performance

	Yang et al scheme	Liu et al scheme	Xu et al scheme	Wu et al scheme
Exponential operations	4	4	4	2
Hash Function	7	7	5	8
Bitwise operations	4	0	3	0
All rounds	3	3	4	3
Security level	DLP	DLP	DLP	DLP

Wu et al. [10] scheme is an efficient authentication scheme for Telecare Medical Information System. In their scheme, they added a new stage named the pre-computing phase. In pre-computing phase, the patient can compute certain utility that require expensive, time-consuming exponential performance and then stores them into the storage device. When these values are required, the patient can extract rapidly from the device thus increase the performance. They declare their scheme is secure and acceptable for small computation devices such as in the Telecare Medical Information System.

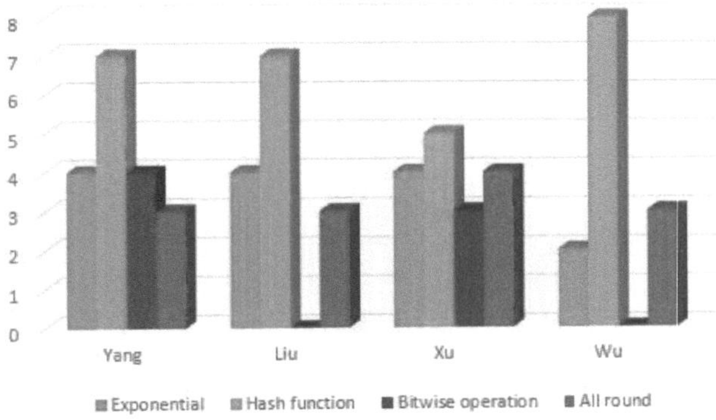

Fig 2.2.1 Comparisons of the performance

Wu et al. [10] strategy suffers from the insider's attack and the impersonation attack. He et al introduce new authentication scheme. Now the performance and the attacks of Xu et al and Wu et al scheme given. The register two strategy [9, 10] need more exponential operations conducting to the require for extra calculation time. He et al [11] proposed scheme only requires 9 hash function operations, one exponential operations.

Xu et al. [9] scheme stand from the impersonation attack. Wu et al. [10] scheme stand from the insider attack. This shows that the he et al scheme is more systematic and suitable in collocating with small power devices for the Telecare Medical Information System.

Xu et al. [10] scheme stand from a password dictionary attack on Lee et al. [8] scheme when the patient's smart card is lost. Pu et al. [31] scheme discuss the same problem of the scheme present by Wang et al. [32]. Latest, Wu et al. [10] scheme present a systematic authentication scheme for Telecare Medical Information System. He et al. [11] scheme could not withstand replay attacks and insider attacks. They initiate a secure authentication scheme for Telecare Medical Information System.

The performance of Wei et al [12] scheme, compare it with Wu et al. [10] scheme and He et al. [11] scheme as shown in Table 4 For benefit, we denote by Ex as experimental and HA as one-way hash function.

As shown in Table 3, pre-computing phase is remove in Wei et al scheme, and the patient does not require to ask the server to produce new pre-computed repeatedly. Wei et al. [12] scheme require some times of experimental operations as Wu et al. [10] scheme. Wu et al. [10] scheme and He et al. [11] schemes, which suffer from offline password guessing attacks when the patient's smart card is lost, Wei et al [12] scheme present a two-factor protocol for authentication, and improve the security of Telecare medical information System.

Table 2.2.3 Comparisons with performance and security parameter

Properties		Schemes		
		Wu et al	He et al	Wei et al
Performance	Computation Cost	2Ex + 8HA	1Ex+ 9HA	2Ex+ 8HA
	Communication rounds	3	3	3
	Pre-computing phase	Yes	Yes	No
Security	Off-line password	No	Yes	No

Modification attack	Yes	Yes	No
Impersonation attack	No	Yes	No
Insider attack	No	Yes	No
Parallel attack	Yes	Yes	No
On-line password	Yes	Yes	No
Replay attack	Yes	Yes	No

Wei et al. [12] introduce that both of Wu et al. [10] scheme and He et al. [11] scheme cannot provide a two-factor protocol for authentication. Wei et al. [12] present an enhance scheme for authentication scheme for Telecare Medical Information System. Zhu et al [13] scheme is vulnerable to an off-line password guessing attack. The security analysis and performance analysis scheme is more suitable for Telecare Medical Information System than Wei et al. [12] scheme.

Wei et al. [12] scheme and its improvement proposed by Zhu et al. [13] scheme fail to reach necessary characteristics for secure patient authentication. Khan et al [14] scheme show that security issue of Wei et al.'s scheme sticks with Zhu et al scheme; like online password guessing attack, user's stolen/lost smart card and replay threat. Khan et al [14] scheme therefore present an authentication scheme for Telecare Medical Information System which protect the confidentiality of messages even if master secret key of telecare medical information system server is compromised.

Table 2.2.4 Comparison of various attacks

Scheme → ↓ Resistance to attacks	Wei et al	Zhu et al	Khan et al	Xu et al	Islam et al	Shehzad et al
Online password guessing attack	Insecure	Insecure	Secure	Secure	Secure	Secure
False login attempts	Insecure	Insecure	Secure	Secure	Secure	Secure
Offline password guessing attack	Insecure	Secure	Secure	Insecure	Secure	Secure
Denial-of-service attack	Insecure	Insecure	Secure	Secure	Secure	Secure

Insider attack	Secure	Secure	Secure	Secure	Secure	Secure
Reply attack	Secure	Secure	Secure	Secure	Secure	Secure
Impersonation attack	Secure	Secure	Secure	Secure	Insecure	Secure
Stolen verifier attack	Secure	Secure	Secure	Secure	Secure	Secure
User anonymity	Insecure	Insecure	Insecure	Secure	Secure	Secure
Two Factor Authentication	Insecure	Secure	Secure	Secure	Secure	Secure
Man-in-middle Attack	Insecure	Insecure	Insecure	Insecure	Secure	Secure
Session key Confidentiality	Secure	NR	Secure	Secure	Secure	Secure

Zhu et al [13] scheme [13] suffers from server impersonation attacks, DoS attacks. Bin et al [15] scheme observe that Zhu et al [13] scheme has unacceptable password change phase. Bin et al [15] scheme is an improve version of Zhu's scheme in which suffer from user anonymous attack.

Bin Muhaya [15] scheme is vulnerable to impersonation attack, does not provide perfect forward secrecy and off-line password guessing attacks. Hamed et al [16] scheme a new two-factor authentication and key agreement scheme apply on elliptic curve cryptosystem. Security analyses of hamed et al [16] scheme does not overcome the problem of Bin Muhaya [15] scheme. Hamed et al scheme is faster than 2.73 times of Bin Muhaya [15] scheme.

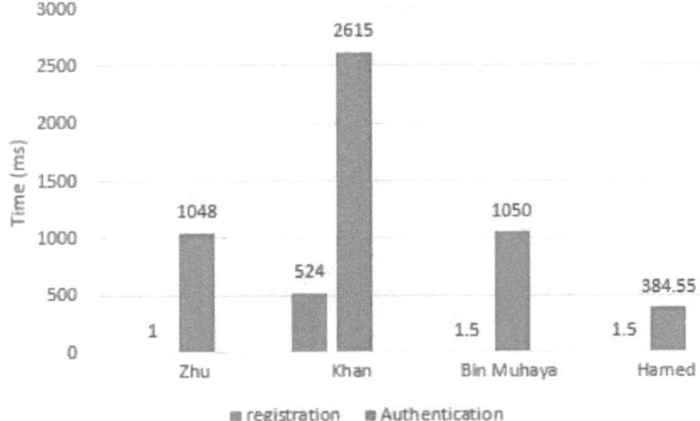

Fig 2.2.2 Running times of different schemes

Lee et al [17] scheme observe that Zhu et al scheme [13] scheme has insecure against parallel attack. Lee et al proposed secure authentication scheme against Zhu et al [13] scheme.to overcome the parallel attack on telecare medical information systems.

Table 5 lists the performance of the other related schemes and lee et al [17] proposed enhanced scheme, where the time T_h for one-way hash function operation, T_{me} for modular exponentiation operation and T_w for modular multiplication operation. The simulation experiments for comparison. lee et al [17] scheme use the intel(R) core(TM)2 Quad CPU Q8200n @ 2.40 GHz with 4.00 GB memory for simulation and the algorithms used are SHA-1 and RSA. The response times, calculated in *millisecond* (*ms*), of the user and server.

Khan et al. [3] examine the security of Wang et al. [35] scheme. Wang et al. [35] scheme does not provide perfect forward secrecy and session key agreement. They scheme also vulnerable to insider attack. Khan et al [3] scheme analyzed the Wang et al [33] scheme has practical issue.

Lee et al [17] scheme has two security problem. Das et al [18] introduce an improvement of their scheme. Lee et al scheme is efficient as compared to Das et al [18] scheme. Further, through the security analysis, Das et al [18] shown that scheme is secure against possible known attacks. In addition to shown that formal security verification using the AVISPA tool to show that Das et al [18] scheme is secure against all attack.

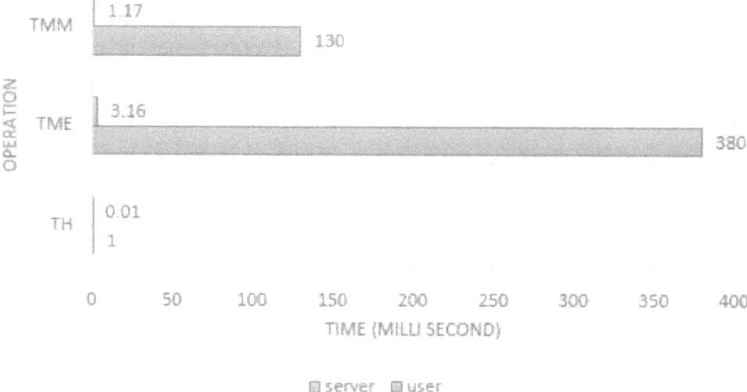

Fig 2.2.3 Execution Time of user side and server side

T_H=One-way hash function operation

T_{ME}=Modular Exponentiation operation

T_{MM}=Modular Multiplication operation

Table 2.2.5 Execution Time of user side and server side

	T_H	T_{ME}	T_{MM}
User	1 ms	380 ms	130 ms
Server	0.01 ms	3.16 ms	1.17 ms

Table 2.2.6 Time operation of different scheme

| Time → | Patient | | | Server | | | Total |
Scheme ↓	T_H	T_{ME}	T_{MM}	T_H	T_{ME}	T_{MM}	Time
Wu et al [10]	4	1	0	4	1	0	387.2 ms
He et al [11]	5	0	0	4	1	0	8.2 ms
Wei et al [12]	5	1	0	5	1	1	389.38 ms
Zhu et al [13]	5	1	0	4	1	0	388.2 ms
Khan et al [14]	5	2	0	5	4	0	777.69 ms
Bin Muhaya et al [15]	3	0	0	12	2	0	9.44 ms
Hamed et al [16]	3	0	0	12	0	6	10.14 ms
Lee et al [17]	5	1	0	5	1	0	388.17 ms
Das et al [18]	9	0	0	14	2	0	15.46 ms
Xu et al [19]	6	0	3	5	0	3	399.56 ms
Islam et al [20]	6	0	2	3	0	1	267.2 ms
Shehad et al [21]	5	0	3	3	0	1	396.2 ms
Li et al [23]	10	0	0	12	0	0	10.12 ms
Ruhul et al [25]	8	0	6	10	0	6	795.12 ms

In 2013, Li et al. [23] analyzed the security of Lee et al [22] chaotic maps based user authentication scheme has security problem. They pointed out that lack of user identity, then proposed an improve scheme. In 2015, Wang et al. [24] introduced that Li scheme [23] suffer from impersonation attack. Latest, Ruhul et al. [25] introduce an elliptic curve cryptography-based scheme for user authentication and key agreement scheme for Telecare Medical Information System environment.

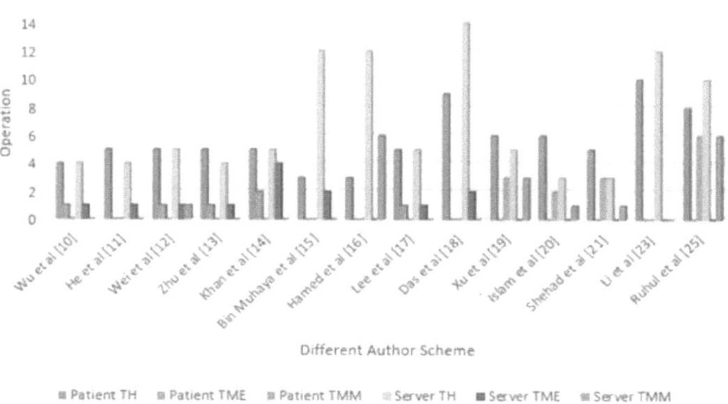

Fig 2.2.4 Time operation of different scheme

3. List of All Attack on Telecare Medical Information System

3.1 Stolen smart card attack

- In stolen smart card attacker will use secret information.

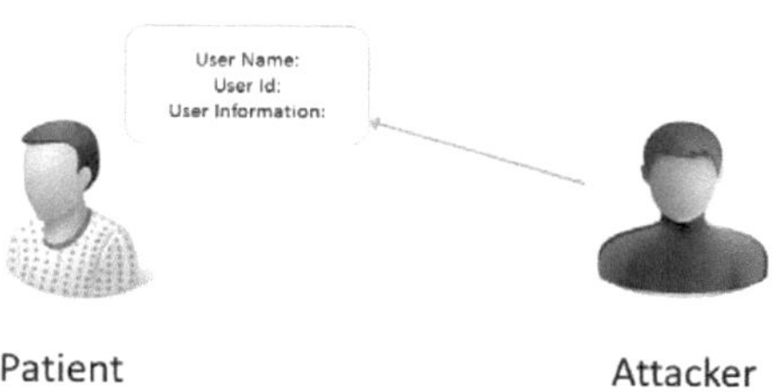

Fig 3.1.1 stolen smart card attacks

3.2 Password guessing attack

- In this attack, attacker can be guessing password online or offline.

Fig 3.2.1 Password guessing attacks

3.3 Insider attack

- In this attack, attacker can work as part of the system of network.

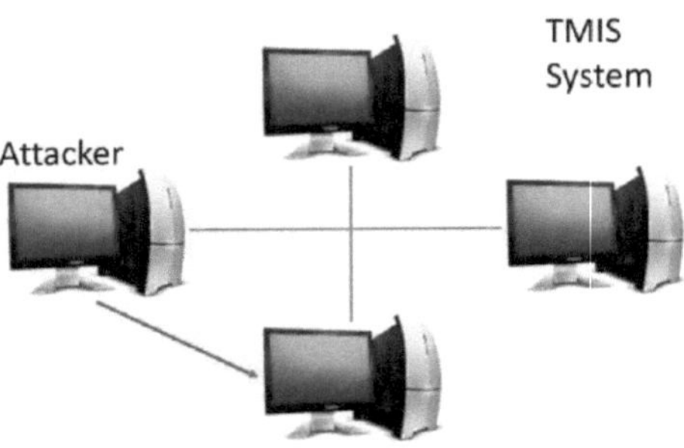

Fig 3.3.1 Insider Attacks

3.4 Reply attack

- In this attack, attacker sends the same message to server.

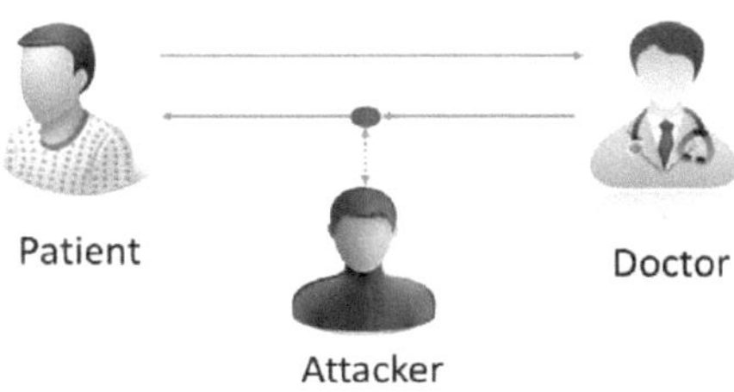

Fig 3.4.1 Reply attacks

3.5 Masquerade attack

- In this attack that patient a fake identity, such as a medical server identity, to gain unauthorized access to attacker through legitimate access identification.

Fig 3.5.1 Masquerade attack

3.6 Stolen Verifier attack

- In stolen verifier, attacker will use the pre-shared secret value to impersonate as patient.

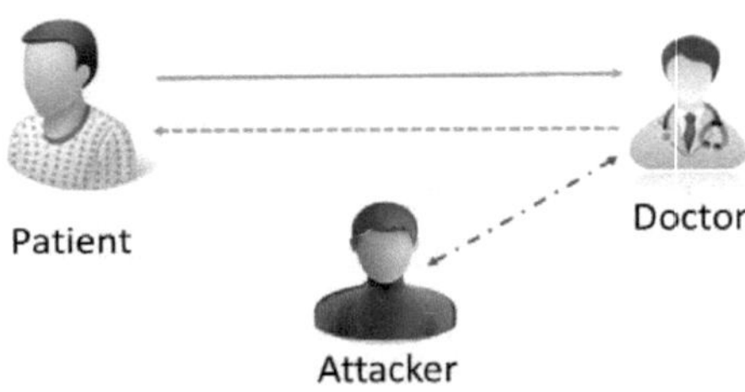

Fig 3.6.1 stolen verifier attacks

3.7 Server Spoofing attack

- There are generally two type DNS spoofing and IP spoofing

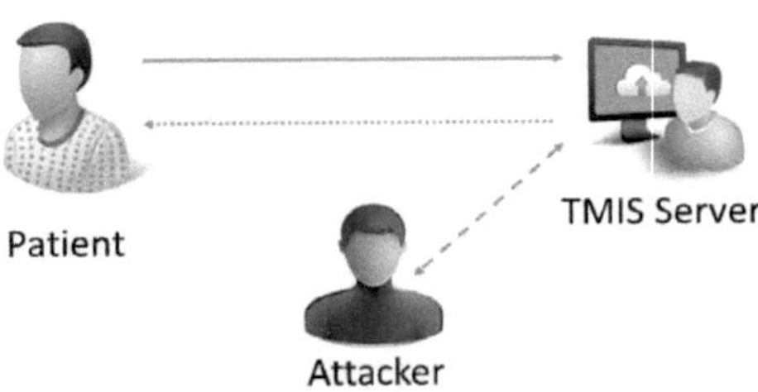

Fig 3.7.1 Server spoofing attack

3.8 Dictionary attack

- In this attacker apply common dictionary work as password.

Fig 3.8.1 Dictionary Attacks

3.9 Parallel session attack

- Attacker will record the value of past session and send it to fool patient or server.

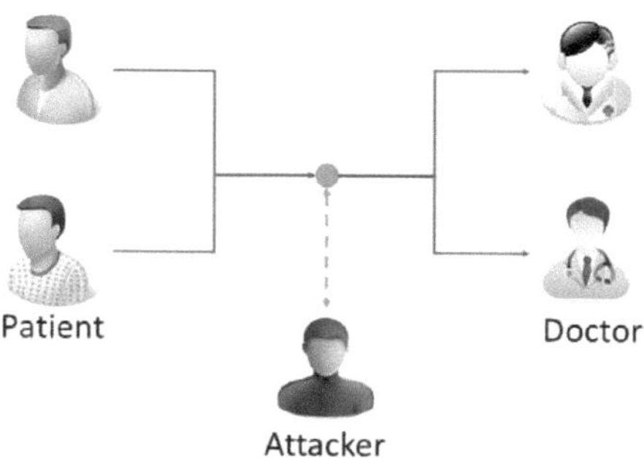

Fig 3.9.1 Parallel session attack

3.10 Modification attack

- Attacker can modify communication message as server

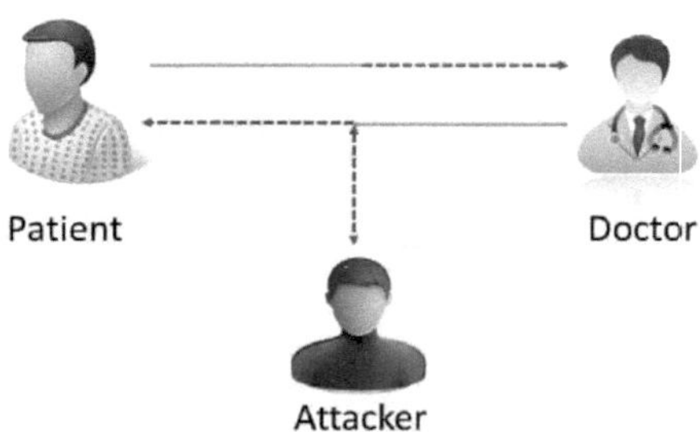

Fig 3.10.1 Modification attack

3.11 Man-in-middle attack

- If attacker can impersonate as initiator and responder, then he can able to do man-in-middle attack.

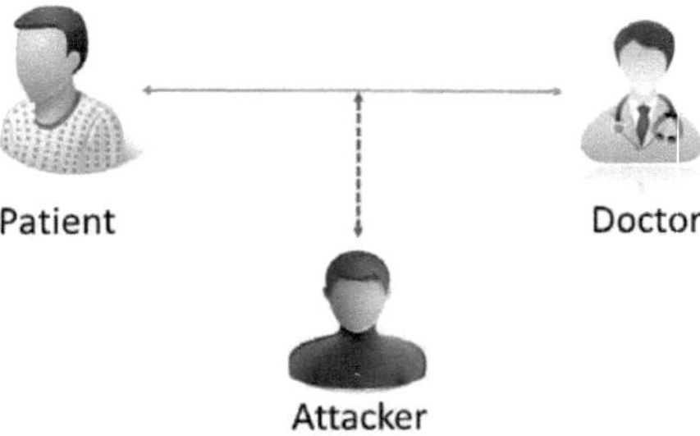

Fig 3.11.1 Man-in-middle attack

3.12 Denial of service attack

- In this attack the attackers attempt to prevent legitimate patient from accessing the service from telecare medical server.

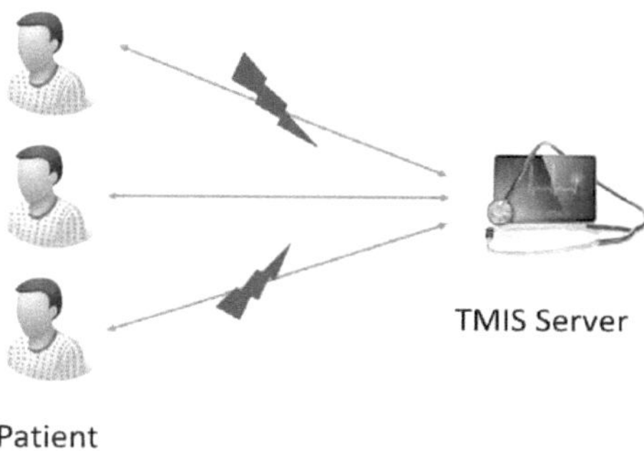

Fig 3.12.1 Denial of service attack

3.13 Session key disclosure attack

- A session key is randomly generated key to ensure the security of a communications session between a patient and telecare medical server.

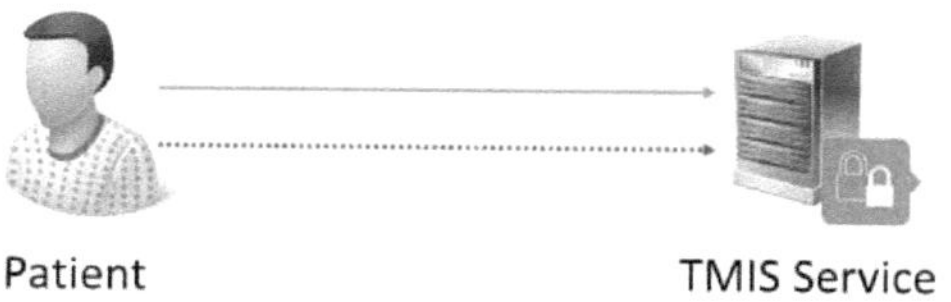

Fig 3.13.1 Session key disclosure attack

- Key escrow is a cryptographic key exchange process in which a key is held in escrow, or stored, by a third party.

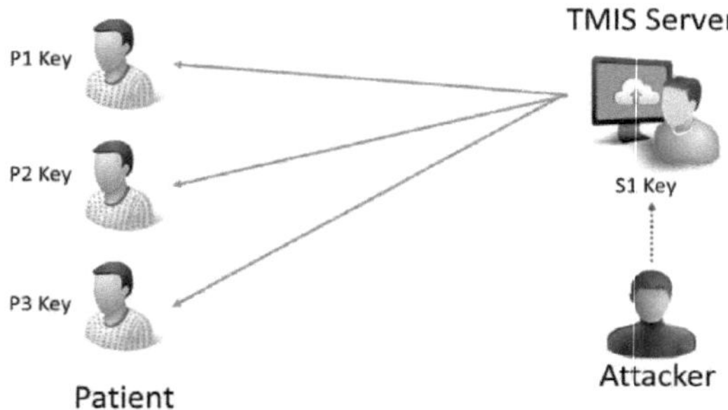

Fig 3.14.1 Key escrow attacks

3.15 Impersonation attack

User Impersonation Attack

- In this attack, attacker can masquerade as sender (or initiator) patient to server.

Server Impersonation Attack

- In this attack, attacker can masquerade as receiver (or responder) server to patient.

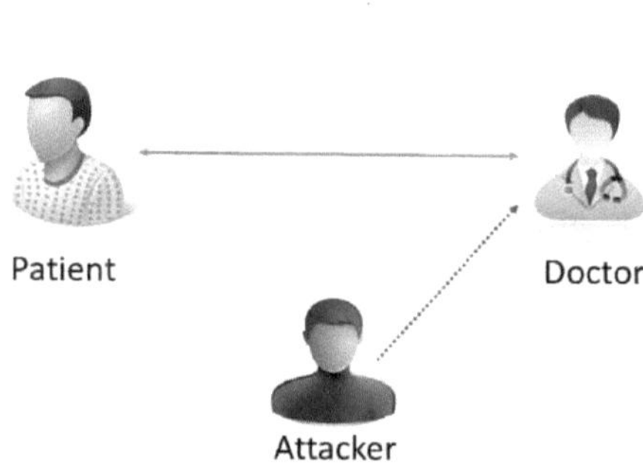

Fig 3.15.1 Impersonation attack

3.16 Forward secrecy attack

- In perfect forward secrecy. Compromise of the long term secrets of three parties cannot able to get the past session keys.

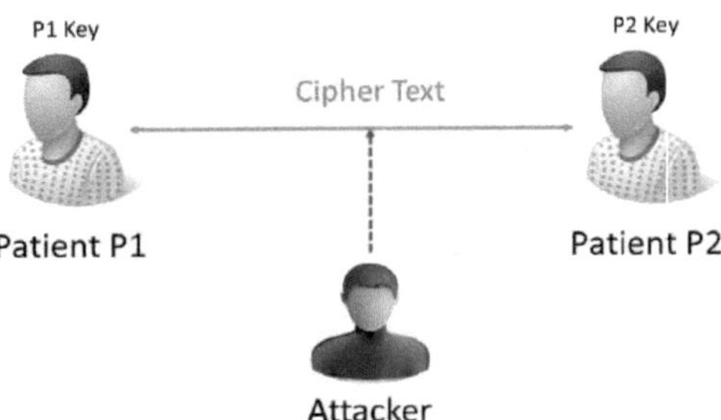

Fig 3.16.1 Forward secrecy attack

3.17 Temporary information attack

- In the known session specific temporary information attack, even if attacker learns the previous session key.

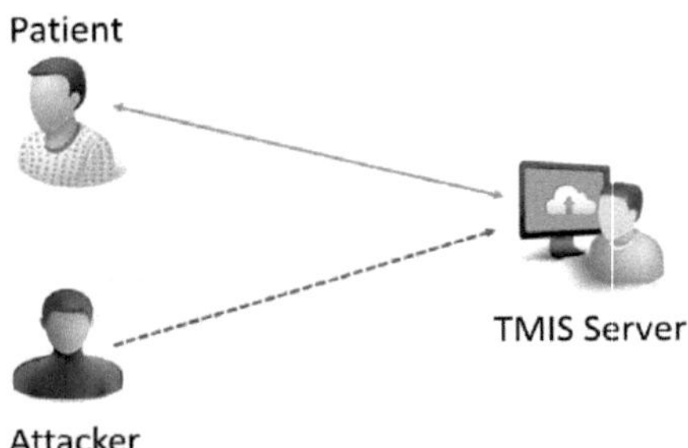

Fig 3.17.1 Temporize information attack

3.18 Survey tree of all attacks

Fig 3.18.1 Survey tree of all attacks

4. Requirement and objective for TMIS

4. 1 Requirements

A practical authentication scheme for Telecare Medical Information System should satisfy the following requirement.

1. A verified User (patient) is allowed to access the TMIS server and then obtain his/her Electronic Medical Record.

2. Mutual authentication and session key agreement could be reached between patients and the TMIS server to stabilize the security of transmitting information.

3. User anonymity should be assured during the communication to protect patient's privacy.

4. Patient can change his/her password freely to get achieve user friendliness.

5. Various common known attacks could be withstanding to ensure security in each session.

6. A low-cost authentication scheme is required for the limited devices.

4.2 Objectives
1. To maintain electronic record of patients' medical history which can be easily accessible.
2. To integrate scattered medical record of patients available at different TMIS service providers.
3. To provide remote medical care to patients at their home via Internet and so on.

5. Conclusion

In this book, we have surveyed all telecare medical information system related paper. Secure password based two factor authentication protocol scheme for all the telecare medicine information system. To withstand the different author schemes which have security weaknesses. The security analysis, shows that author scheme is secure against attacks including impersonation attack, replay attack, online password guessing, off line password guessing, session key security, parallel session attack, user anonymity, stolen smart card attack, man-in-the-middle attack. Automated validation of internet Security Protocols Applications (AVISPA) OFMC back-end for the formal security verification, secure against passive and active attacks, including the replay and man-in the-middle attacks.

References

1. L. Lamport, "Password authentication with insecure communication," *Commun. ACM*, vol. 24, no. 11, pp. 770–772, (1981). Doi: 10.1145/358790.358797
2. M. Hwang and L. Li, "A new remote user authentication scheme using smart cards," *IEEE Trans. Consum. Electron.*, no. 2, pp. 2–4, 2000. Doi: 10.1109/30.826377
3. M. K. Khan, S. K. Kim, and K. Alghathbar, "Cryptanalysis and security enhancement of a 'more efficient & secure dynamic ID-based remote user authentication scheme,'" *Comput. Commun.*, vol. 34, no. 3, pp. 305–309, 2011. doi:10.1016/j.comcom.2010.02.011
4. H.-M. Chen, J.-W. Lo, and C.-K. Yeh, "An Efficient and Secure Dynamic ID-based Authentication Scheme for Telecare Medical Information Systems," *J. Med. Syst.*, vol. 36, no. 6, pp. 3907–3915, (2012). doi:10.1007/s10916-012-9862-y.
5. Q. Jiang, J. Ma, Z. Ma, and G. Li, "A privacy enhanced authentication scheme for telecare medical information systems," *J. Med. Syst.*, vol. 37, no. 1, (2013). doi:10.1007/s10916-012-9897-0.
6. T. Cao and J. Zhai, "Improved dynamic ID-based authentication scheme for telecare medical information systems," *J. Med. Syst.*, vol. 37, no. 2, 2013. Doi: 10.1007/s10916-012-9912-5
7. Q. Xie, J. Zhang, and N. Dong, "Robust anonymous authentication scheme for telecare medical information systems," *J. Med. Syst.*, vol. 37, no. 2, 2013. Doi: 10.1007/s10916-012-9911-6
8. N. Y. Lee and Y. C. Chiu, "Improved remote authentication scheme with smart card," *Comput. Stand. Interfaces*, vol. 27, no. 2, pp. 177–180, 2005. Doi: doi:10.1016/j.csi.2004.06.001
9. J. Xu, W. T. Zhu, and D. G. Feng, "An improved smart card based password authentication scheme with provable security," *Comput. Stand. Interfaces*, vol. 31, no. 4, pp. 723–728, (2009). doi:10.1016/j.csi.2008.09.006
10. Z. Y. Wu, Y. C. Lee, F. Lai, H. C. Lee, and Y. Chung, "A secure authentication scheme for telecare medicine information systems," *J. Med. Syst.*, vol. 36, no. 3, pp. 1529–1535, (2010). doi:10.1007/s10916-010-9614-9.
11. H. Debiao, C. Jianhua, and Z. Rui, "A More Secure Authentication Scheme for Telecare Medicine Information Systems," *J. Med. Syst.*, vol. 36, no. 3, pp. 1989–1995, (2011). doi:10.1007/s10916-011-9658-5
12. J. Wei, X. Hu, and W. Liu, "An Improved Authentication Scheme for Telecare Medicine Information Systems," *J. Med. Syst.*, vol. 36, no. 6, pp. 3597–3604, (2012). doi:10.1007/s10916-012-9835-1.
13. Z. Zhu, "An Efficient Authentication Scheme for Telecare Medicine Information Systems," *J. Med. Syst.*, vol. 36, no. 6, pp. 3833–3838, (2012). doi:10.1007/s10916-012-9856-9.
14. M. K. Khan and S. Kumari, "An Authentication Scheme for Secure Access to Healthcare Services," *J. Med. Syst.*, vol. 37, no. 4, p. 9954, (2013). doi:10.1007/s10916-013-9952-5.
15. F. T. Bin Muhaya, "Cryptanalysis and security enhancement of Zhu's authentication scheme for Telecare medicine information system," *Secur. Comm. Networks*, vol. 8, no. 2, pp. 71–81, 2014. DOI: 10.1002/sec.967
16. H. Arshad, V. Teymoori, M. Nikooghadam, and H. Abbassi, "On the Security of a Two-Factor Authentication and Key Agreement Scheme for Telecare Medicine Information Systems," *J. Med. Syst.*, vol. 39, no. 8, p. 76, (2015). doi: 10.1007/s10916-015-0259-6
17. T. F. Lee and C. M. Liu, "A secure smart-card based authentication and key agreement scheme for telecare medicine information systems," *J. Med. Syst.*, vol. 37, no. 3, (2013). doi: 10.1007/s10916-013-9933-8
18. A. K. Das and B. Bruhadeshwar, "An Improved and Effective Secure Password-Based Authentication and Key Agreement Scheme Using Smart Cards for the Telecare Medicine Information System," *J. Med. Syst.*, vol. 37, no. 5, (2013). Doi: 10.1007/s10916-013-9969-9
19. X. Xu, P. Zhu, Q. Wen, Z. Jin, H. Zhang, and L. He, "A Secure and Efficient Authentication and Key Agreement Scheme Based on ECC for Telecare Medicine Information Systems," *J. Med. Syst.*, vol. 38, no. 1, (2013). Doi:10.1007/s10916-013-9994-8
20. S. H. Islam and M. K. Khan, "Cryptanalysis and Improvement of Authentication and Key Agreement Protocols for Telecare Medicine Information Systems," *J. Med. Syst.*, vol. 38, no. 10, p. 135, (2014). doi:10.1007/s10916-014-0135-9.
21. S. A. Chaudhry, H. Naqvi, T. Shon, M. Sher, and M. S. Farash, "Cryptanalysis and Improvement of an Improved Two Factor Authentication Protocol for Telecare Medical Information Systems," *J. Med. Syst.*, vol. 39, no. 6, (2015). doi:10.1007/s10916-015-0244-0
22. Lee, C. C., Hsu, C.W., Lai, Y. M., and Vasilakos, A., An enhance mobile-healthcare emergency system based on extended chaotic maps. J. Med. Syst. 37(5):1–12, 2013. DOI 10.1007/s10916-013-9973-0
23. Li, X., Niu, J., Khan, M. K., and Liao, J., An enhanced smart card based remote user password authentication scheme. J. Netw. Comput. Appl. 36(5):1365–1371, 2013. DOI 10.1007/s10916-014-0077-2
24. Wang, Z., Huo, Z., and Shi, W., A dynamic identity based authentication scheme using chaotic maps for telecare medicine information systems. Journal of medical systems 39(1):1–8, 2015. DOI 10.1007/s10916-014-0158-2

25. Ruhul A, SK H Islam, G. P. Biswas, M. K. Khan, Neeraj K., An Efficient and Practical Smart Card Based Anonymity Preserving User Authentication Scheme for TMIS using Elliptic Curve Cryptography. J Med Syst. 39(180) 2015. DOI 10.1007/s10916-015-0351-y

26. Lee, W. B., and Lee, C. D., A cryptographic key management solution for HIPAA privacy/security regulations. IEEE Trans. Inf. Technol. Biomed. 12(1):34–41, 2008. Doi: 10.1109/TITB.2007.906101

27. Lambrinoudakis, C., and Gritzalis, S., Managing medical and insurance information through a smart-card-based information system. J. Med. Syst. 24(4):213–234, 2000. Doi: 10.1023/A:1005549330655

28. Diffie, W., and Hellman, M., New directions in cryptology. IEEE Trans. Inf. Theory 22(6):644–654, 1976. Doi: 10.1109/TIT.1976.1055638

29. Stallings, W., Cryptography and network security: Principal and practices. 4th Edition. Prentice Hall, 2005. ISBN 13: 978-0-13-609704-4

30. ElGamal, T., A public-key cryptosystem and a signature scheme based on discrete logarithms. IEEE Trans. Inf. Theory IT-31 (4):469–472, 1985. Doi: 10.1007/3-540-39568-7_2

31. Pu, Q., Wang, J., and Zhao, R. Y., Strong authentication scheme for telecare medicine information systems. *J. Med. Syst.* (2011). doi:10.1007/s10916-011-9735-9.

32. Wang, R. C., Juang, W. S., and Lei, C. L., Provably secure and efficient identification and key agreement protocol with user anonymity. *J. Comput. Syst. Sci.* 2010. doi:10.1016/j.jcss.2010.07.004

33. Liao, E., Lee, C.C., and Hwang, M.S., A password authentication scheme over insecure networks. *J. Comput. Syst. Sci.*, 72(4):727–740, 2006. doi:10.1016/j.jcss.2005.10.001

34. Das, M. L., Saxena, A., and Gulati, V. P., A dynamic ID-based remote user authentication scheme. *IEEE Trans. Consum. Electron.* 50(2):629–631, 2004. Doi: arXiv:0712.2235

35. Wang, Y.-Y., Liu, J.-Y., Xiao, F.-X., and Dan, J., A more efficient and secure dynamic ID-based remote user authentication scheme. *Comput. Commun.* 32(4):583–585, 2009. doi:10.1016/j.comcom.2008.11.008

Acronyms

Abbreviations	Denotation
AVISPA	Automated Validation of Internet Security Protocols and Applications
CPU	Center Processing Unit
DLP	Discrete logarithm problem
EMR	Electronic medical record
HHC	**Home healthcare agency**
OFMC	On-the-Fly Model-Checker
RSA	Rivest Shamir Adleman
TMIS	Telecare medical information system

Mr. Kishan Makadia received Bachelor of Engineering in Computer science engineering from Sanjaybhai Rajguru College of Engineering, Rajkot under Gujarat Technological University (GTU), Ahmedabad, India in 2014. He is currently pursing Master of Engineering in Computer Engineering from Marwadi Education Foundation Group of Institution (MEFGI), Rajkot, India under GTU. He is interested in Research on Cryptography, Information Security, Two factor authentication, and Telecare medical information system. He is life member of Cryptology Research Society of India (CRSI), Kolkata, India.

Mr. Sunil Vithlani received Bachelor of Engineering in Computer Engineering from V.V.P. Engineering college, Rajkot under Saurashtra University, Rajkot, India in 2011. He has completed M.E. from Marwadi education Foundation, Rajkot in 2013. Currently he is working in the Department of Computer Engineering at Marwadi Education Foundation, Rajkot since 2014. he is interested in Research on Cryptography and Ad-hoc Networks.

Dr. Nishant Doshi is a faculty in the Department of Computer Engineering at Marwadi Education Foundation, Rajkot since 2014. His main research interests include algorithms, cryptography and remote user authentication, information protection in general. He has completed M. Tech from DA-IICT, Gandhinagar in 2009 and Ph.D. from NIT Surat in 2014. Along with active researcher, he is Editor-in-Chief of journals like IJCES, IJECEE, IJME, IJMES, and IJSCE. He is rewarded as *Young Scientist* from Venus International Foundation in year 2015.

YOUR KNOWLEDGE HAS VALUE

- We will publish your bachelor's and
  master's thesis, essays and papers

- Your own eBook and book -
  sold worldwide in all relevant shops

- Earn money with each sale

Upload your text at www.GRIN.com
and publish for free